Edison Twp. Free Public Library
340 Plainfield Ave.
Edison, New Jersey 08817

My First Science Words

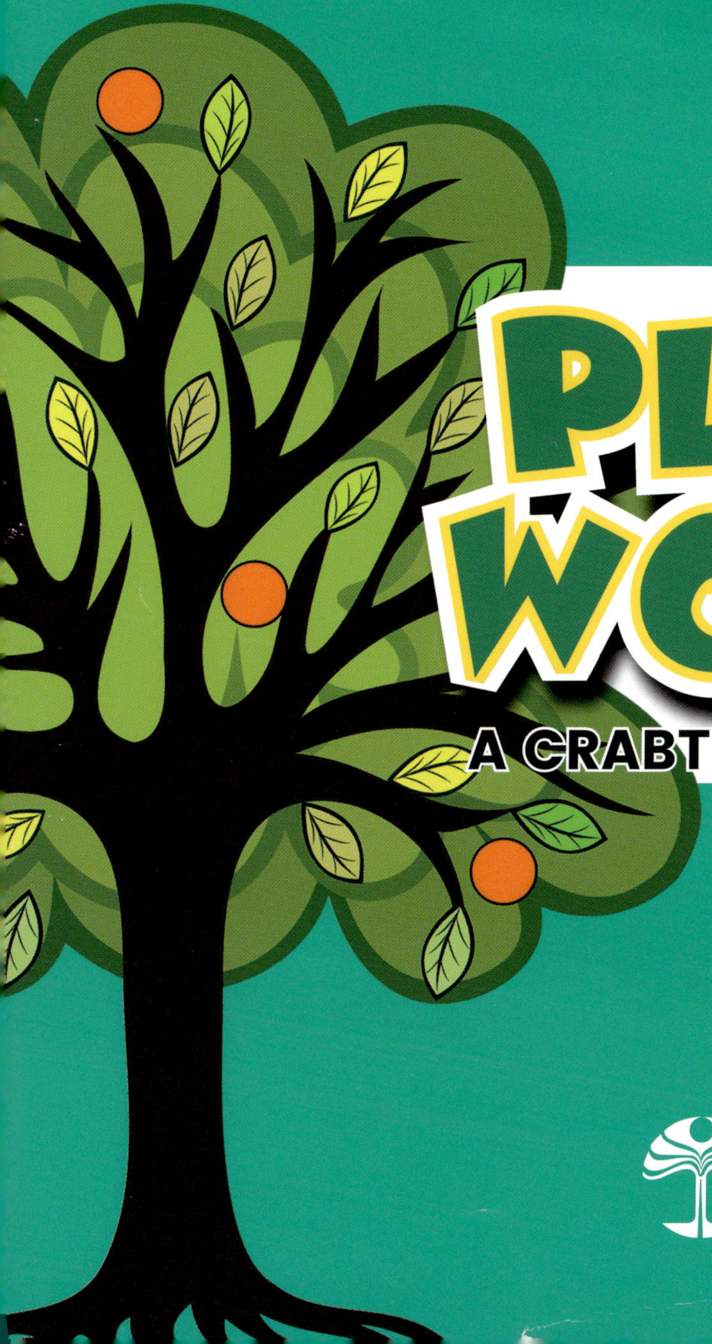

PLANT WORDS

A CRABTREE SEEDLINGS BOOK

Taylor Farley

plants
(PLANTS)

seeds
(SEEDZ)

soil

(SOYL)

roots
(ROOTS)

stem
(STEM)

leaf
(LEEF)

flower
(FLOU-ur)

petal
(PET-uhl)

fruit
(FROOT)

sunlight
(SUHN-lite)

water
(WAW-tur)

Glossary

flower (FLOU-ur): A flower is the colorful part of the plant.

fruit (FROOT): The fruit is the fleshy, juicy part of the plant. Fruits contain seeds.

leaf (LEEF): A leaf grows out from a plant stem, twig, or branch. Leaves use sunlight to make food for the plant.

petal (PET-uhl): A petal is one of the outer parts of a flower.

plants (PLANTS): Plants are living things. Many plants grow from seeds and have stems, roots, leaves, and flowers.

roots (ROOTS): Roots are the parts of a plant that grow underground. Water travels up the roots to the plant stem.

seeds (SEEDZ): Seeds are plant parts that can grow new plants.

soil (SOYL): Soil is dirt or earth. Many kinds of plants grow in soil.

stem (STEM): A stem is the long, main part of the plant. Leaves and flowers grow out from the stem.

sunlight (SUHN-lite): Sunlight is the light from the Sun. Many plants use sunlight to make their food.

water (WAW-tur): Water is a liquid which has no color. Plants and animals need water to live.

School-to-Home Support for Caregivers and Teachers

Crabtree Seedlings books help children grow by letting them practice reading. Here are a few guiding questions to help the reader build his or her comprehension skills. Possible answers are included.

Before Reading

- **What do I think this book is about?** I think this book is about plants. It might tell us about how plants grow.
- **What do I want to learn about this topic?** I want to learn about the different parts of a plant.

During Reading

- **I wonder why...** I wonder why the watermelon grows close to the ground.
- **What have I learned so far?** I have learned that roots, a stem, a leaf, and a petal are parts of a plant.

After Reading

- **What details did I learn about this topic?** I learned that plants need sunlight and water.
- **Write down unfamiliar words and ask questions to help understand their meaning.** I see the word *soil* on page 6 and the word *fruit* on page 18. The other vocabulary words are found on pages 22 and 23.

Library and Archives Canada Cataloguing in Publication

Title: Plant words / Taylor Farley.
Names: Farley, Taylor, author.
Description: Series statement: My first science words | "A Crabtree seedlings book". | Previously published in electronic format by Blue Door Education in 2020.
Identifiers: Canadiana 20200384899 | ISBN 9781427130457 (hardcover) | ISBN 9781427130501 (softcover)
Subjects: LCSH: Plants—Terminology—Juvenile literature. | LCSH: Botany—Terminology—Juvenile literature.
Classification: LCC QK49 .F37 2021 | DDC j580—dc23

Library of Congress Cataloging-in-Publication Data

Names: Farley, Taylor, author.
Title: Plant words / Taylor Farley.
Description: New York : Crabtree Publishing, 2021. | Series: My first science words : a crabtree seedlings book | Audience: Ages 4-6 | Audience: Grades K-1 | Summary: "This book builds beginning vocabulary about plant science. Extremely helpful for elementary science preparation, eight words combine with a visual depiction so readers can see what the word means"-- Provided by publisher.
Identifiers: LCCN 2020049658 | ISBN 9781427130457 (hardcover) | ISBN 9781427130501 (paperback)
Subjects: LCSH: Botany--Juvenile literature. | Plants--Juvenile literature.
Classification: LCC QK49 .F26 2021 | DDC 580--dc23
LC record available at https://lccn.loc.gov/2020049658

Crabtree Publishing Company

www.crabtreebooks.com   1–800–387–7650

e-book ISBN 978-1-949354-09-6

Print book version produced jointly with Blue Door Education in 2021

Content produced and published by Blue Door Publishing LLC dba Blue Door Education, Melbourne Beach FL USA. Copyright Blue Door Publishing LLC. All rights reserved. No part of this book may be reproduced or utilized in any form or by any means, electronic or mechanical including photocopying, recording, or by any information storage and retrieval system without permission in writing from the publisher.

Written by Taylor Farley
Production coordinator and Prepress technician: Samara Parent
Print coordinator: Katherine Berti

Printed in the U.S.A./012021/CG20201102

Photo credits: page 1 ©istock.com/ vicvic13 ; photo page 3 © shutterstock.com/Artorn Thongtukit; illustrations pages 4 and 6 © shutterstock.com/ aekikuis; photo page 5 © shutterstock.com/Lisa F. Youn, photo page 7 © shutterstock.com/domnitsky, illustrations pages 8, 10, 12 © shutterstock.com/Glasscage; photo page 9 © shutterstock.com/Kobkit Chamchod; photo page 11 © shutterstock.com/Lotus Images; illustration page 13 © shutterstock.com/Designua, photo page 13 © shutterstock.com/Romolo Tavani; illustration page 14 © shutterstock.com/BlueRingMedia; photo page 15 © shutterstock.com/Naruto_Japan123, illustration page 16 © shutterstock.com/sersupervector, photo page 17 © shutterstock.com/plampy; illustration page 18 © shutterstock.com/Mr Escape, photo page 19 © shutterstock.com/tchara; illustration © 31moonlight31, page 21 © shutterstock.com/PHOTO JUNCTION ; page 22 ilustration ©Kolonko, page 23 © shutterstock.com/dugdax Back cover ©istock.com/ vspn24

Published in Canada
Crabtree Publishing
616 Welland Ave.
St. Catharines, Ontario
L2M 5V6

Published in the United States
Crabtree Publishing
347 Fifth Ave.
Suite 1402-145
New York, NY 10016

Published in the United Kingdom
Crabtree Publishing
Maritime House
Basin Road North, Hove
BN41 1WR

Published in Australia
Crabtree Publishing
Unit 3 – 5 Currumbin Court
Capalaba
QLD 4157

Edison Twp. Free Public Library
340 Plainfield Ave.
Edison, New Jersey 08817